Leonhard Euler

A Man to be Reckoned With

Text by Andreas K. Heyne
and Alice K. Heyne

Illustrations by
Elena S. Pini

English Translation by
Alice K. Heyne and Tahu Matheson

 Birkhäuser

Leonhard Euler – the most important facts:

Euler lived in the 18th century, a time marked by the conflict between absolute rule and the objectives of the enlightenment; objectives that were rooted in the idea that rational man would overcome outdated thoughts and beliefs. Artistically speaking, he lived during the late baroque or rococo period. His contemporaries were Bach and Lessing, and later Mozart and Schiller.

He achieved extraordinary things in many different fields of mathematics, as well as excelling in numerous other subjects. The importance of his work cannot be overestimated, and his name should be placed alongside those of Newton and Einstein. Even today, his theories and formulas are still used by engineers, architects and statisticians around the world.

He was, however, not a neglected genius. After leaving his hometown, Basel in Switzerland, never to return, he lived in Berlin and St. Petersburg, and worked with the greatest minds of his generation. His numerous publications were the bestsellers of the time.

Due to technical reasons it has not been possible to redraw the scenes. In some places, where the German text had been integrated into the picture, some text has been left in the original language. However, where possible, a translation in footnotes has been added.

Text: Andreas K. Heyne and Alice K. Heyne
Illustrations: Elena S. Pini

© 2007 Bernoulli-Euler-Gesellschaft, Basel

This book is published under the imprint Birkhäuser, www.birkhauser-science.com, by the registered company Springer Nature Switzerland AG.

The registered company address is: Gewerbestrasse 11, 6330 Cham, Switzerland

ISBN 978-3-7643-8332-9

Leonh. Euler

Леонгардъ Эйлеръ

$$\pi = 4as - \frac{3}{a} + \frac{3}{6aa}$$

$$\text{et } \pi = 3,166666$$

$$aas = 4,6$$

$$\frac{1}{5} + \frac{1}{8} = 0,576$$

$$e = 23,14069$$

$$\pi = 3,14159$$

2

4

* Algebra textbook from 1553

Margreth, I think it's time to send the boy to school in Basel.

Already!

Be a good boy and obey your grandma!

May God protect you, my son, and study hard!

At last the questioning is over!

And so...

BASEL 1½ MILES

in good weather

Finally...

6

Good day, I'm Leo.

Hi grandma! Looks as if you're stuck with me now.

An entire room for me alone!

And my own desk!

Grammar School*
"auf Burg"

The lessons are not...

...and the consecutio temporum...

...to everybody's taste...

...and they fly better...

...the rounder and heavier they are.

You can easily calculate the trajectory if you...

Euler, this is your last warning!

...and don't offer much variety.

de bello gallico Caesar

EULER

* Humanistisches Gymnasium (HG)

But sometimes even the dullest lessons can stir the imagination:

Could you summarize the building of the bridge for us again?

...and if Caesar had built an arched bridge and had used fewer cross-struts, ...

...he would have had to fell about 500 trees fewer and could have conquered the rest of Gaul as well!

The school-days end with...

Finally the boring lessons are over!

...a bonfire.

But...

...his joy was premature, ...

Gentlemen, Marcus Aurelius...

...the boredom continues...

Those Romans are getting on my nerves!

And how!

...and continues...

However, sooner or later...

If in the isoperimetric problem $ddy = 0$ is taken, one has

$$y = \int \frac{(x \pm c)\, dx}{\sqrt{a^2 + (x \pm c)^2}}$$

and with that the tautochrone is determined.

Notizen zu Bernoulli Variationsrechnung *

...something grabs you...

Shhh!

Sorry, but I can't...

...bear to listen to that any more. I have it at home all day!

Why, who are you?

Johann, Johann Bernoulli**, and the chap out the front is my dad, it's a real pain, I tell you!

* Notes on Bernoulli's Calculus of Variations
** The Bernoullis were a famous family of mathematicians from Basel.

...and doesn't let you go...

Mathematica est omnis divisa in partes duas...

* Heute: Leonh. Euler zum Thema Arithmetik „u. Geometrie"

Perfect! This will be my Masters Thesis: On Descartes' Vortices and Newton's Gravitation.

Action at a distance: Wouldn't that be magic?...

...so the planets are dragged along by aether vortices.

Phew – everything went well!

After passing the exam...

Johann, I did it!

We need to celebrate this!

Let's go to my place!

Wow, they're all books about mathematics!

Come on, I'll show you my new tin soldiers!

Just a second...

* Today: Leonhard Euler on the subject: "Arithmetics and Geometry"

*When I was a child, I spoke as a child... (1 Corinthians 13:11)

12

I only got the consolation prize in the Paris competition - they probably assumed someone from Switzerland wouldn't know anything about ships...

...and I didn't get the Physics professorship either...

But Niklaus and Daniel got a job with the Czar in St.Petersburg in the new academy!

How I envy them! It's supposed to be a paradise for scholars.

Listen, those two can surely get you...

...a job there, and Dad could also intercede for you - what else are fathers there for?

Indeed, in St.Petersburg...

My husband, the Czar, has founded the academy but we still don't have the number of professors we need. Don't the gentlemen Bernoulli know some other intelligent people?

Of course! Euler! He always argued with Dad. In fact he was the only one who understood Dad.

Euler? Can he take over the professorship for medicine?

Well, as a matter of fact he is...

Shh! Shut up will you! We need a fourth man to play cards!

...the...

...best professor of medicine far and wide!

And so...

MARZIPAN

Buddenbrook & Söhne

This is where you have to wait for the ship to St.Petersburg, Mr.Euler.

ABFAHRT nach ST.PETERSBURG

Krug

Lübeck

Every available moment is used...

Das komplette Medizin-Studium in 30 Tagen...*

PSCHYREMBEL

MATERIA MEDICA

...purposefully...   ...well,

Anatomie für Dummies**

DIAGNOSIS

...almost every one!

* A Complete Guide to Medicine in 30 Days   ** Anatomy for Dummies

If we had turbines we would go much faster – must calculate this sometime...

Finally...

Hello, Christian!

ST. PETERSBURG HARBOUR CONTROL

IVAN FOR KING

Good thing you came to pick me up, there's a lot going on here!

Yes, we've got some serious problems at the moment: The Czarina died, and total chaos broke out. It's good that you can stay with Daniel for the moment.

Perhaps I should just go back home...

That's out of the question! The other guys from Basel are pining for you already...

You mean they're in desperate need of my mathematical knowledge here?

...that too. But first of all...

...we need a fourth man to play cards!

Ah, Euler, there you are at last! Since Niklaus died we've managed to teach...

...Goldbach the card game. How's Dad? Still as despotic as ever?

This is Jakob, by the way.

Jakob Hermann. Ych kumm au us Baasel.*

I have no time to play cards at the moment, my medical knowledge is still inadequate!

*I'm from Basel too.

...a few months later...

Johann has come to St.Petersburg because he thinks that I wouldn't be able to find my way back to...

Yum! Swiss chocolate!

...Switzerland without him! The Swiss community has organised a great farewell-party for me tonight. Are you coming?

I'd love to!

Don't you think, if mathematics is pure nature, it should be found in everything? For example in music? Or in colours? My father is a painter you know.

Fascinating thought, Miss Gsell, would you like to discuss this further sometime?

She would...

A long and happy life to our newly-weds Leonhard and Katharina!

Apart from contracting...

...a severe illness, which cost him his right eye, Euler began a period...

...of most fruitful creativity.

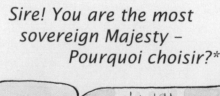

Sire! You are the most sovereign Majesty – Pourquoi choisir?*

Superb, Voltaire! Put together a list of the top scholars - I want the best for my academy. Offer them a better salary than they have now!

As Monseigneur wishes. And the Silesians?

We'll send them an offer too, but without a salary increase, haha!

To Berlin? I don't know...

Of course we'll go! Here you never know what is going to happen to the government, and the fires everywhere ... besides, the salary in Berlin is much better!

I don't know. It seems much too uncertain to me...

The case is decided as in every good marriage...

* Why choose?

But already the welcome doesn't live up to expectations...

Berlin Academy of Science

Scheduled completion: Anytime in the next 10 years...

(perhaps)

Der Keenig? Nee, der is in Schleesien. Nee, ick weess ooch nich, wann der wieder-kommt...*

...wenn iberhoopt.**

Meanwhile at the Silesian border...

...and I want the baths in Glatz. I...

...need them urgently: My doctor has prescribed me to take a cure.

No, I have no time for the academy at the moment! War must come first! Tell Euler he should find something to occupy himself with...

*The king? No, he's in Silesia. No, I don't know when he is coming back either...   **...if he comes back, that is...

No problem for Euler...

Special edition!

Special edition!

The war is over! Silesia is ours! Special edition!

Well, now he can finally start his academy. That is, if nothing else intervenes...

Special edition!

At the same time Empress Maria Theresia in Vienna...

Now we've lost Silesia to the damn Prussians! Well then, we'll take Bavaria. What do you think, Franzl?

Whatever you want, darling, whatever you want.

She wants...

I've no time for the academy as long as this Austrian hussy allies herself with...

...all my enemies. Tell Euler he should find something to occupy himself with...

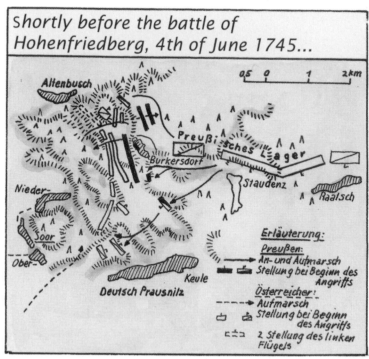

Shortly before the battle of Hohenfriedberg, 4th of June 1745...

Perhaps we should attack at night?

Damn it! It's that Euler again!

He really must occupy himself with something more important...

...with artillery for example!

* "The knight's tour", a mathematical problem based on the way a knight moves on the chessboard, was something that Euler studied intensively.

y

...a signed copy of my book* chronicling my expedition to the pole, I brought some exemplaires at a special price...

Since he has proved...

...that the poles are flat...

...his brain too has lost some of its sharpness.

Even at court more peaceful activities prevail...

I have composed a melody, see if you can't make something decent of it, Bach!

* Original title: Sur la figure de la terre

Your Majesty, ...

...it would be an honour to present you with a small musical offering.

Bernoulli has died. Now they are offering you his position in Basel. But with the salary they're proposing, it's out of the question! For God's sake, will you leave my wine-glasses in peace!

Meanwhile the academy runs smoothly...

...and I'm still certain zat zis letter by Leibniz is a fake. And if...

...somebody wants another signed copy of my book, I brought some exemplaires at a special price...

...only the appreciation of some scientific works leaves a little to be desired...

...my theory of the Moon, Majesty...

Can't you write something useful for a change, Euler? How about turbines and water power for example? The Oder-Havel Canal has to be levelled and the Oder Marshes have to be drained. Furthermore, I have plenty of dams to build and...

As Your Majesty wishes...

...here I have a real challenge for you, Euler: Someone has to install the fountains in the park for my next party...

* Friedrich, we shudder at you! (alludes to a quote from Goethe, Faust, 1,2: Margaret: "Henry! I shudder now to look on thee!")
** 4 years of war – how much longer?   *** Stop that – I'm virtually a Russian!

34

Прекратите!
Я его знаю!*

Professor Euler! It was you who tried to teach me algebra at the military academy! Actually, I never understood it properly, but I will always be grateful that you didn't flunk me...

Отнесите барахло немедля назад! Этот человек — табу!**

...the famous Euler? I've read your Theory of Music: Just marvellous...

But slowly more peaceful times return...

May I present myself: Heinrich, margrave of Brandenburg-Schwedt. Aren't you...

Thank you very much.

With your pedagogic talents, you might even succeed in teaching my lazy daughter something! Doesn't have to be mathematics, that's not for girls, just general stuff if you know what I mean...
Perhaps you could write to her sometime?

*Stop it – I know him!   **Put that stuff back! This man is taboo!

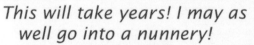

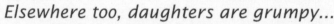

*Published as "Letters of Euler on different Subects in Natural Philosophy Addressed to a German Princess".

Now that we've got Glatz again we will go there to take a cure, won't we, Franzl?

Whatever you want, darling...

In Prussia, however, the mood is much less relaxed after the battle of Landshut...

I know that since Maupertuis left you've been running the academy, but I won't give you the directorship...

...why do you need a raise anyway? I've been told that you are doing very well from the lottery!

Signed:

Friedrich

And finally peace...

Thank God for Peter!

Special edition! Czarina Elisabeth dead! Surprising peace offer from the new Czar!

...but not everywhere...

That's an impudence! ...

...Now that we are at peace, he really could give you the position! Did he write...

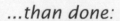

Easier said...

...therefore I would like to request my dismissal...

...than done:

...therefore, I would **again** like to request my dismissal **urgently**...

Resignation 1st Version

Versions of 3rd attempt to resign

New attempt to resign: Version 27

L. Euler

...and if another request for dismissal arrives from Euler, throw it away!

But powerful friends sometimes work miracles...

...and write to Friedrich, telling him that if he doesn't let Euler go, I shall be forced to reconsider my husband's peace treaty!

...and so...

All right! Go!
    Signed: Friedrich.
P.S. Of course, as your son Christoph is enlisted in my army, you will have to leave him behind – nobody gets out of there quickly (he he he)

The packing takes time, but finally...

During their journey, they stop in Warsaw where the new king, Stanislaus II, gives Euler a brilliant reception.

...and tell the charming Catherine that I miss her very much, especially the discussions long into the night...

I will tell her, Your Majesty.

The journey continues by ship...

I want my rocking-horse!

That's on the second ship with the other luggage.

...but on the way...

Blast...

...it! Now I'll have to write the whole Integral Calculus again!

Grandpa, my rocking-horse!

41

But somewhere else the "elegant literature" is truly appreciated...

The reception by Catherine the Great is a triumph...

I have arranged the positions...

...for your sons, and I will advance you the money for a house, as long, my dear...

...Euler, as you make your calculations for US! I will pamper you so much, ahem, I mean...

...I will shower you with so many honours, that you will never want...

...to go back to the Prussians.

You may count on me, Your Majesty.

We can count on him, but can we reckon with him?
ha ha ha

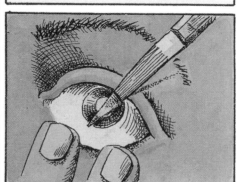

Peace and quiet return to Euler's everday life, even if it's not always easy: It is no longer possible to postpone his eye operation...

dr-stroke-stroke equals dt times d-to-the-power-of-4-z divided by dx-dy squared-dt

Stop – not so fast! Was that divided by dx squared-dy-dt?

You'd better get used to it, I'm wondering if I'll be able to see anything at all in the future; the operation seems to have made everything worse.

...occasionally things even become rather fiery...

Leonhard! – for heaven's sake come down!

My formulas, my calculations!

You Russian fool, I need my documents!

If you go back in there, all you will need is a coffin!

I can't let a fellow-countryman be burned to death, can I? Ych kumm nämlig au us Baasel!*

However, excepting little interruptions, Euler continues to live in St. Petersburg until the end of his days...

Dear Euler, I will compensate you for your house. By the way, ...

*I'm from Basel too, you know!

...perhaps it's time you wrote something practical...

...I have a little problem involving a bridge over the Neva river...

Oh no! After that Königsberg episode*, I never want to hear the word "bridge" again.

Of course Euler solves the problem of the span of the bridge and many, many, many, more...

Special edition! First balloon flight a success!

...because his curiosity...

Sensation! Man can fly!

I'll take one, please!

...never ceases...

It's true! The brothers Montgolfier have flown in their hot air balloon! Incredible!

How big...

* One of the many mathematical problems Euler solved, was the so-called "problem of the seven bridges of Königsberg".

...until the present day have been occupied with the publication of his complete works...

...this may still take a while!

By the way: One really can't rely on authors these days. It's true that they've done a pretty good job describing and drawing the life of Leonhard Euler, but occasionally, when it comes to historical precision, they haven't always been particularly accurate. Some anachronisms have found their way into the pictures and the text – things which wouldn't have existed during Euler's lifetime, or at least would not have been known in Europe. Perhaps the reader might enjoy tracking down a mistake or two? Euler probably would have ...

Leonhard Euler (1707–1783)
Life and Work

Youth 1707–1727

Leonhard Euler was born on the 15th of April, 1707, in Basel, Switzerland. The exact position of the house of his birth is unknown. In 1708 Euler's father became the parson of Riehen, and the family moved into the parsonage there. In 1713 Euler went to the classical secondary school in Basel. While he was there, he was able to stay with his grandmother during the week.

At the (then usual) age of 13, Euler enrolled in philosophy at the University of Basel. He completed his studies three years later, in 1723, and received his master's. In 1724, aged 17, he held his first public lecture, comparing the philosophies of Descartes and Newton. Then, respecting the wishes of his father, Euler enrolled in theology.

However, Euler was seen much more often in Johann Bernoulli's lectures on geometry, arithmetic, and astronomy. Bernoulli's youngest son, Johann II, had received his degree at the same time as Euler, and it was through him that Euler was able to meet and converse with the famous scholar. His insatiable curiosity was somewhat satiated by the great man who answered many of his questions and recommended the seminal works of mathematics to him. In due course, he gave up theology, and turned his full attention to mathematics.

Aged 18, Euler wrote his first treatise, solving a particularly difficult problem that Bernoulli had set for the best mathematicians of the time. In 1736 he submitted a thesis on sound, as an application for a position at the University of Basel, but was rejected.

St. Petersburg 1727–1741

In 1725, the two elder Bernoulli brothers, Daniel and Nikolaus, were already well paid lecturers in St. Petersburg. They (along with the secretary of the academy, Christian Goldbach) secured an invitation for Euler to become a junior lecturer in physiology in St. Petersburg, one of the great centres of intellectual learning at the time. On the 5th of April 1727, Euler left Basel never to return.

When Euler arrived in St. Petersburg, the city was in a state of political unrest. A few days earlier, the Czarina, Catherine I, had died, and the succession of the throne was uncertain. As a result, many intellectuals left the Academy. Euler began his work in this atmosphere of great political turmoil. 3 years later, in 1730, the 15 year old Czar, Peter II, died. His cousin, Anna Ivanovna, succeeded him and set about enforcing the rule of law immediately. In 1731 Euler became professor of physics and a member of the Academy, and in 1733 professor of mathematics. The superior professor's salary enabled him to marry Katharina Gsell, the daughter of a Swiss painter living in St. Petersburg. The happy union produced 13 children, but only 3 sons and 2 daughters reached adulthood.

Until his marriage, Euler had lived with Daniel Bernoulli. He then bought a house on the Vassilievski Island, and his first son, Johann Albrecht, was born there.

Euler took on more and more duties at the Academy. In 1735 he became supervisor of the geography department, a member of the Committee for Measures and Weights, supervised secondary school and cadet school exams, and held lectures there. In spite of all that, he was incredibly productive. In the Academy journal from 1736, 11 of 13 contributions were from Euler (the other two were from Daniel Bernoulli).

Most important works of the first St. Petersburg period*:
1729	Infinite Series
1730	Two important works on geodesic lines
1731	Attempt at a new theory of music
1734	Investigations of Euler's constant γ

1735	On the sums of series of reciprocals
	On planetary orbits
	Solution of the «Königsberg Bridge problem»
1736	Mechanics (2 volumes)
1737	Continued fractions
1738	Successful answer to the Paris Academy's competition problem about the theory of tides
1738/40	The art of reckoning (2 volumes)

In 1740 Anna Ivanovna died, and the working conditions for foreigners deteriorated drastically. Due to the political instability and the constant threat of fires in St. Petersburg, Euler accepted an offer from Frederick II and moved his family to Berlin on the 19th of June 1741.

Berlin 1741–1766

When Euler arrived in Berlin, Frederick II was away fighting a war in Silesia. Euler busied himself setting up the observatory. It wasn't until two years later, in 1743, that Frederick could turn his attention to the Academy, which in 1746 moved into rooms in the palace. Euler became the head of mathematics, and Maupertuis became president. In 1742, Euler bought a house near the planned site for the academy. In 1748, he refused the offer to become Johann Bernoulli's successor in Basel – the salary was insufficient.

Euler did not have a particularly good relationship with Frederick II, who thought him too unworldly, but the king did consult him on many technical issues. From 1753 right through to the end of the Seven Years' War (1756–63), Euler effectively ran the academy. Nevertheless, at the end of the war, he was not given the directorship. Frederick II offered the position to d'Alembert, who declined, suggesting Euler. Frederick refused. In 1766, Euler handed in his resignation.

Most important works of the Berlin period*:
1742	Proof of "Fermat's little theorem"
1744	Calculus of variations
	Theory of the motions of planets and comets
1744–53	10 works on the "principle of least action" by Maupertuis
1745	New principles of gunnery
1746	Thoughts on the elements of bodies (with this, Euler intervened in the quarrel about monads – he sided against Leibniz, and thus offended Wolff and some of his colleagues at the Academy)
1747	Defence of divine revelation against the objections of the atheists
	On the ambiguity of logarithms
1748	Introduction to the analysis of the infinite (2 volumes)
	Differential calculus (2 volumes)
1749	Naval science (2 volumes)
1750	Contemplations on space and time
	Theorem on changes of momentum
	Polyhedron formula
1753	Works on turbines and hydrodynamics
	First theory of the motion of the Moon (Euler received part of the prize money of the English Longitude Competition for this, which he had to share with clockmaker Harrison and the widow of the astronomer Mayer
1756	Further development of the calculus of variation by Lagrange
1758	On the motion of a solid rotating round a mobile axis
1762	Achromatic lenses
1765	Theory of the motion of rigid bodies

---

* The dates indicate when the works were written or presented, rather than published: there can often be a gap of several years between the different dates.

St. Petersburg 1766–1783

On the 1st of June, 1766, the 59 year old Euler and his house-hold of 18 (including domestic staff) left Berlin. Catherine II, who had been reigning in St. Petersburg since 1762, prepared a triumphant reception for him on his return. Euler acquired a house at the quay on the river Neva near the Academy. He received a princely salary, and secured good positions for his sons: Johann Albrecht, for instance, became the professor of experimental physics.

After a sudden illness Euler became almost blind. This did not, however, affect his productivity, half his works stem from after he became blind. A fellow countryman from Basel, Nikolaus Fuss, became his assistant.

In May 1771, a fire ravaged the Vassilievski Island, destroying 500 houses, including Euler's. Another countryman, named Grimm, pulled him from the flames at the last minute.

The star physician Baron von Wenzel performed an operation on Euler's eye, which was temporarily a success. However, after an infection he lost his sight altogether.

In 1773, Euler's wife of 40 years died. 3 years later he married her half-sister, Salome Gsell. Between 1780 and 1781 both his daughters died, but numerous grandchildren rallied round him.

In his last years Euler withdrew more and more from the Academy, due to his age and because of differences with the director. In 1774 he and his son left the committee. However, in 1783 the intelligent princess Dashkova became the director of the Academy. For her inaugural speech she invited Euler to be the guest of honour. Presumably this was his last appearance at the Academy. He died on the 17th of September after a stroke, aged 76.

Most important works of the second St. Petersburg period:

1768	Letters to a German Princess on various topics in physics and philosophy (3 volumes, written 1760–62)
1768–70	Foundations of Integral Calculus (3 volumes)
1769–71	Optics (3 volumes)
1770	Complete instruction in algebra (2 volumes)
1772	Second theory of the motion of the Moon
1773	Complete theory of the construction and steering of ships
1775	General principle of conservation of angular momentum
1777	Determination of coefficients in trigonometric series

Euler was buried in the Lutheran graveyard on Vassilievski Island, and a marble statue of him was erected in the Academy. Euler's descendants in Russia were respected citizens, mostly officials or engineers.

The Academy of St. Petersburg was occupied with the publication of his articles until 1862. Statistically speaking, Euler must have made an important discovery every week. There are still thousands of pages of unpublished manuscripts in the archive of the Academy. The Euler Commission in Basel has also been, and still is, in the process of publishing his complete works in 75 volumes.

P. S. In the information pavilion of the 'Viaduc de Millau', Europe's highest motorway bridge, opened in 2004, a bridge that effortlessly spans the valley between Clermont-Ferrand and Montpellier in the shape of an elegant curve, there is a dedication to Leonhard Euler.

And with good reason: The essential calculations on the wind currents, vibrations, and stability rely on Euler's formulas.

Crowned heads around Leonhard Euler

Frederick II the Great (1712–1786)

The Prussian king lived with a constant inner conflict between his artistic talents and spiritual and religious tolerance on one side, and the Prussian severity and will to power on the other. During the Silesian Wars 1740/42 and 1744/45 he annexed Silesia. In 1756 he began the Seven Years' War (again for Silesia and against Maria Theresia of Austria). He was victorious, and Prussia became one of the great powers in Europe. The Academy of Science he established in Berlin was amongst the most liberal in Europe.

Maria Theresia (1717–1780)

The Archduchess of Austria and Queen of Hungary and Bohemia was the greatest opponent of FREDERICK II. She lost Silesia and Parma-Piacenza in the course of the Silesian Wars. She was one of the most capable rulers of the Habsburg Monarchy, and is still venerated today. She abolished torture and witch trials, founded the public primary school and ran the Habsburg multicultural state excellently. Her daughter Maria-Antonia (Marie-Antoinette) married Louis XVI of France and was beheaded in the course of the French Revolution.

Peter I the Great (1672–1725)

The most important of all Russian Czars had only one aim in life: To make Russia a great European power. In 1697/98 he travelled incognito to Holland to study ship-building. He founded the city of St. Petersburg and made it his capital – strategically, it was excellently placed between Lake Ladoga and the Baltic Sea. In the Northern War against Sweden he gained a decisive victory. Russia became, and has stayed, the most important Baltic power.

Catherine I (1684–1727)

The peasant girl became the mistress of Prince Menshikov, in whose household she met Peter I in 1703. Only after she had given birth to three of his children would the Czar marry her, and pronounced her Czarina in 1724, which effectively made her co-ruler. After his death in 1725, she ruled the empire and encouraged many great scholars – including Euler – to come to the Academy of Science founded by Peter. Unfortunately she died unexpectedly, shortly before Euler arrived in St. Petersburg.

Peter II (1715–1730)

Peter was still a child when, in 1727, he succeeded his mother Catherine I on the throne. However, in reality the empire was ruled by Prince Menshikov, the former lover of his mother. When Peter died of smallpox, his aunt Anna Ivanovna succeeded him.

Anna Ivanovna (1693–1740)

Following the death of Peter II in 1730, the daughter of Ivan V and niece of Peter the Great was made Empress by the Russian high nobility. She re-established autocracy and seriously influenced the architecture of St. Petersburg (for example by her star-shaped street system). As she had no son she proclaimed her grand nephew, Ivan VI, her successor.

Elisabeth (1709–1762)

The daughter of Peter the Great and Catherine I overthrew the government of the child-king Ivan VI in 1741. Due to her sex she had been left out of the succession. As she had no children she proclaimed her nephew, Peter III, her successor. In 1745 she married him to Catherine II (the Great). In 1755 she founded the University of Moscow and in 1757 the Academy of Fine Arts in St. Petersburg.

Catherine II the Great (1729–1796)

Born as Sophie Friederike Auguste von Anhalt-Zerbst in Stettin (Prussia), she led a revolt against her husband, Peter III, in 1762 and proclaimed herself Empress. This highly intelligent and educated woman had as many military conquests as she had lovers. Catherine II considered herself as "philosopher on the throne", and was a great patron of the arts. She thought very highly of Euler and offered him a well paid position at the Academy of Science.

Stanislaw II August Poniatowski (1732–1798)

In 1764 the son of a politician and member of the regional parliament was elected (the last) King of Poland – not least through the support of his former lover Catherine II. In 1795 he had to abdicate and returned to St. Petersburg, where he died.

Further famous contemporaries of Leonhard Euler

(in the order of their year of birth and without the least claim to completeness)

Sir Isaac Newton (1643–1727)

The English physicist, mathematician, astronomer, alchemist, and natural philosopher was the founder of classical theoretical physics. He formulated – among many other things – the three laws of motion of mechanics, the law of universal gravitation, and the theory of emission of light as corpuscles refracted by their acceleration toward a denser medium. Independently of Leibniz he developed differential and integral calculus. In 1703 he was made President of the Royal Society.

Gottfried Wilhelm Leibniz (1646–1716)

The polymath invented the binary system (the system on which the entire computer technique is based). He invented – simultaneously but independently of Newton – infinitesimal calculus, introduced the mathematical term 'function', and constructed the first calculating machine. His theory of monads (a type of "mental atom") as the ultimate element of the universe was controversial.

Johann I Bernoulli (the Elder) (1667–1748)

The Swiss mathematician from Basel occupied himself particularly with series and differential calculus. He was especially fascinated by the definition and examination of curves emerging from questions in mechanics like the brachistochrone (the curve between two points that is traveled in the least time by a body that starts at the first point with zero speed and passes down along the curve to the second point, under the action of constant gravity and ignoring friction).

Christian Wolff (1679–1754)

The German philosopher began his career as professor of mathematics and natural philosophy in Halle and became a member of the Berlin Academy. He had to leave Prussia due to accusations of atheism, and proceeded to Marburg (Hesse-Kassel), from whose university he had received an invitation. There Lomonosov was one of his students. Wolff systematized parts of the philosophy of Leibniz and defined a great part of the (still valid) philosophical terminology. Catherine the Great made him professor with a lifelong pension. Finally Frederick the Great recalled him to Halle, where he became chancellor of the university.

Vitus Jonassen Bering (1681–1741)

The Danish seafarer and explorer of Asia became an officer in the Russian Navy in 1703. In 1728 he reached the easternmost point of the Asian continent and the Bering strait, which was consequently named after him.

Johann Sebastian Bach (1685–1750)

The most important musician of the Baroque period, composed the Musical Offering (BWV 1079) – a set of variations on a theme by the king – after a challenge by Frederick the Great. It consists of three fugues, nine canons and a trio sonata. It is probable that Bach met Euler at court.

Christian Goldbach (1690–1764)

The Prussian lawyer from Königsberg became the first secretary of the newly opened St. Petersburg Academy of Sciences, and later became tutor to the Czar Peter II. From 1742 on he figured as a civil servant at the Russian Foreign Ministry. Nowadays, he is best remembered for Goldbach's conjecture, which he formulated in 1742 in one of his numerous letters to Euler.

Voltaire, real name François-Marie Arouet (1694–1778)

The fame of the most important philosopher of the Enlightenment was so great that the 18th century was called "the century of Voltaire". From 1750 to 1753 he stayed at the court of Frederick the Great in Potsdam. He got into a quarrel with several scholars, including Euler, about the authenticity of a letter by Leibniz. In response he wrote a satirical essay about Maupertuis. Frederick was so furious, that Voltaire had to leave Berlin. Voltaire carried on an extensive correspondence with almost all the important personalities of Europe. Furthermore, he worked for the Encyclopédie of Diderot und d'Alembert.

Johann Joachim Quantz (1697–1773)

The flautist, flute teacher and composer at the court of Frederick II composed about 300 flute concertos and 200 chamber music works, which Frederick himself played.

Pierre Louis Moreau de Maupertuis (1698–1759)

As a member of the French Académie des Sciences, he confirmed, in 1736, on behalf of the French King Louis XV, Newton's thesis on the oblate shape of the earth, by measuring the length of a degree of the meridian in Lapland. In 1746 he became – at the behest of Frederick II – president of the Royal Academy of Sciences in Berlin and developed his "principle of least action" to calculate mechanical motion. Some scientists wanted to ascribe the principle to Leibniz, which led to serious controversy within the Academy. As a result Maupertuis resigned and went to Basel, where he stayed as a guest of his friend Johann II Bernoulli until he died.

Daniel Bernoulli (1700–1782)

The son of Johann I Bernoulli studied medicine in Basel. During an educational stay in Italy he published his first treatises on the so-called Riccati equation, on the probabilities in a card game, and on the efflux of liquids out of a vessel. In 1725 he was invited, along with his brother Nicolaus, to the Academy of St. Petersburg, where he wrote his chief work, the Hydrodynamica. In 1733 he returned to Basel and took over the chair for anatomy and botany, and in 1750 the one for physics. He was one of Euler's close friends.

Benjamin Franklin (1706–1790)

The most important leader of the Enlightenment in the United States also published discoveries and theories regarding electricity – and, incidentally, invented the lightning rod, the Franklin stove, the catheter, the swim fins, the glass harmonica, and bifocals.

Carl von Linné (1707–1778)

The Swedish doctor and naturalist was one of the founders of the Royal Swedish Academy of Sciences in Stockholm in 1739. He created the classification system for plants and animals which we still use today.

Albrecht von Haller (1708–1777)
The Swiss anatomist, physiologist, poet and literary critic worked from 1729 on as a doctor in Berne and later became head of the library. From 1736 to 1753 he was professor of anatomy, surgery and botany at the University of Göttingen. Later he returned to Berne and became Rathausamtmann, schools inspector and head of the orphanage, and finally even the director of some salt mines. In particular his 8 volume work on human physiology made him world famous. In addition he published philosophical works and poems and wrote numerous literary reviews.

Mikhail Vasilevic Lomonosov (1711–1765)
The Russian poet and naturalist is considered to be the reformer of the Russian language. From 1745 on he taught chemistry at the University of St. Petersburg but was also very productive as a poet. He participated in the founding of the Moscow University, and became rector of the St. Petersburg University in 1760. He suggested the wave theory of light and wrote several letters to Euler, in one of which he expresses his theory of the conservation of matter for the first time. Later he set up a glass factory and produced glass mosaics.

Jean-Jacques Rousseau (1712–1778)
The Swiss-French philosopher, writer and composer was by means of his socially critical writings the most important forerunner of the French Revolution.

Alexis Claude Clairaut (1713–1765)
In 1736, the French mathematician, geometer and physicist took part, together with Maupertuis, in an expedition to Lapland. His report about the determination of the shape of the earth became a classic. He published important astronomical works and wrote new textbooks on algebra and geometry.

Denis Diderot (1713–1784)
The French writer and philosopher was the editor-in-chief and most important author of the Encyclopédie ou Dictionnaire raisonné des sciences, des arts et des métiers, on which over 160 scientists collaborated. Many of his works were forbidden by the state and the church, and were published only after the French Revolution. He was partly funded by Catherine II.

Jean Le Rond d'Alembert (1717–1783)
The illegitimate son of a French chevalier studied law and medicine, finally mathematics and physics. He wrote numerous articles for the Encyclopédie, and also edited it along with Diderot.

Adam Smith (1723–1790)
This Scottish philosopher and political economist is thought to be the founder of classical economics.

Immanuel Kant (1724–1804)
The founder of modern philosophy and author of the Critique of Pure Reason, was professor of logic and metaphysics in Königsberg and also published works on mathematics, physics and cosmology.

Giacomo Casanova (1725–1798)
The Venetian playboy travelled as diplomat throughout Europe, and had access to many important personalities such as Voltaire or Frederick II. Apart from his gallant adventures he found the time to publish his memoirs, as well as historical, mathematical and satirical writings.

James Cook (1728–1779)
The greatest of all discoverers undertook three circumnavigations of the globe, and explored Australia, New Zealand, the South Pacific and Alaska. In 1770 Cook claimed Australia for the British Crown.

Gotthold Ephraim Lessing (1729–1781)
The great German poet of the Enlightenment became dramatic adviser at the newly founded German National Theatre in Hamburg in 1767. His dramatic piece Nathan der Weise is considered the ultimate vindication for religious tolerance.

George Washington (1732–1799)
The former tobacco farmer fought as an officer in Virginia, and led America's Continental Army to victory over Britain in the American Revolutionary War. Later, he became the first president of the United States (1789–1797).

Joseph-Louis Lagrange (1736–1813)
The Italian-French astronomer and mathematician published works about differential equations and the calculus of variations. In 1766 he became, on Euler's recommendation, director of the Academy of Sciences in Berlin. In doing so he succeeded Euler (who, however, had never held this office officially). Under Napoleon, Lagrange was proclaimed count and senator of France.

James Watt (1736–1819)
In 1765, this Scottish inventor developed the first working steam engine, which was later used as driving power in textile factories. With this development the Industrial Revolution began. The unit of electrical power, watt (volt x ampere), is named after him.

Frederick William Herschel (1738–1822)
This German-British astronomer, composer and designer of telescopes discovered the planet Uranus in 1781.

Georg Christoph Lichtenberg (1742–1799)
The first German professor for experimental physics became famous, as a writer, for the aphorisms found in his waste books, which were published posthumously.

Antoine Laurent de Lavoisier (1743–1794)
Due to his Elementary Treatise, he became the father of modern chemistry. In the course of the French Revolution, Lavoisier was accused as a blackmailer and beheaded in 1794.

Johann Wolfgang von Goethe (1749–1832)
The greatest of all German poets and dramatists was also an enthusiastic naturalist and anatomist, as is proved by his many scientific works, such as his Theory of Colours.

Wolfgang Amadeus Mozart (1756–1791)
During the span of his short life, the composer from Salzburg created works of eternal beauty in almost all musical forms.

Friedrich von Schiller (1759–1805)
German's most important poet and dramatist after Goethe was also an excellent historian and wrote – amongst others – important works about the Thirty Years' War.